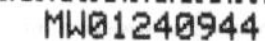

MATHEMATICS:

INAUGURAL ADDRESS

OF

CHARLES DAVIES, LL. D.,

PROFESSOR OF MATHEMATICS IN COLUMBIA COLLEGE,

ON

The Nature, Language, and Uses

OF

MATHEMATICAL SCIENCE,

February 11th, 1858.

NEW YORK:

WYNKOOP, HALLENBECK & THOMAS, PRINTERS,

No. 113 FULTON STREET.

1858.

MATHEMATICS:

INAUGURAL ADDRESS

OF

CHARLES DAVIES, LL. D.,

PROFESSOR OF MATHEMATICS IN COLUMBIA COLLEGE,

ON

The Nature, Language, and Uses

OF

MATHEMATICAL SCIENCE,

February 11th, 1858.

NEW YORK:

WYNKOOP, HALLENBECK & THOMAS, PRINTERS,

No. 113 FULTON STREET.

1858.

ADDRESS.

The first, and surely the most difficult duty assigned to me by the Board of Trustees, is that of explaining to a popular audience the nature of Mathematical Science—the forms of its language—its uses as a means of mental training and development—its value as the true basis of the practical—the sources of knowledge which it opens to the mind and the place which it should occupy in a justly balanced system of Collegiate instruction.

The term Mathematics, as used by the ancients, embraced every known Science and was also applicable to all other branches of Knowledge. Subsequently, it was restricted to those more difficult subjects which require continuous attention, severe study, patient investigation and exact reasoning; and such subjects were called Disciplinal, or Mathematical.

Mathematics, as a science, is conversant about the laws of Numbers and Space. The two abstract quantities, Number and Space, are the only subjects of Mathematical Science. The laws which are evolved in the processes employed in searching out the elements of these abstract quantities, in discussing their relations, and in framing a proper language by means of which these relations can be recorded and a knowledge of them communicated, constitute the Science of Mathematics. The faculties of the mind chiefly employed in the cultivation of this Science are simply, the apprehension, the judgment and the reasoning faculty.

The term quantity, applicable both to number and space, embraces but eight classes of units: 1st, Abstract Units; 2d, Units of Currency; 3d, Units of Length; 4th, Units of Surface; 5th, Units of Volume; 6th, Units of Weight; 7th, Units of Time; and 8th, Units of Angular Measure.

The laws which make up the Science of Mathematics are established in a series of logical propositions, deduced from a few self-evident notions of these unities, which are all referred to number and space. All the definitions and axioms, and all the truths deduced from them, by processes of reasoning, are therefore traceable to these two sources.

In mathematics, names imply the existence of the things which they name, and the definitions of those names express attributes of the things. Hence, all definitions do, in fact, rest on the intuitive inference that things corresponding to the words defined have a conceivable existence as subjects of thought, and do, or may have, proximatively, an actual existence. Every definition of this class is a tacit assumption of some proposition, which is expressed by means of the definition, and which gives to such definition its importance.

The axioms of Geometry are intuitive inductions; that is, they are perfectly conceived by a single process of the mind, without the intervention of other ideas, the moment the facts on which they depend are apprehended. When we say, "A whole is equal to the sum of all its parts," or, "A whole is greater than any of its parts," the mind immediately refers to a single thing, divided into parts; it then compares the whole thing with all its parts, or the whole thing with some of its parts; and then infers, by a process of generalization, that what is true of one thing and its parts is also true of every other thing and its parts: so that these axioms, however self-evident, are still generalized propositions, and so far of the inductive kind, that, independently of experience, they would not present themselves spontaneously to the mind.

The pure mathematics being based on definitions and axioms, as premises, all its truths are established by processes of deductive reasoning; hence, it is purely a deductive science. If all the connections between the minor and major premises were obvious to the senses, or as evident as the truth, "A whole is equal to the sum of all its parts," there would be no necessity for trains of reasoning, and deductive science would not exist. Trains of reasoning are necessary for extending the definitions and axioms to new cases; and there is no logical test of truth, in the whole range of mathematical science, except in the conformity of the conclusions to the definitions and axioms, or to such known principles as may have been established from them.

Language is a collection of all the signs of thought by means of which we express our ideas and their relations. The language of mathematics is mixed. It is composed partly of symbols, which have a precise and known signification, and partly of words borrowed from our common language. The symbols, although arbitrary marks, are, nevertheless, entirely general in their signification, as signs and instruments of thought, and when the sense in which they are used is once fixed, by definition, they always retain the same meaning throughout the same process. The meaning of the words taken from our common vocabulary is often modified and sometimes entirely changed, when transferred to the language of science. They are then used in a particular sense and are said to have a technical signification.

There are three principal forms, or dialects of the

Mathematical language: the language of Number, of which the elementary symbols are the ten figures: the language of Geometry, of which the elements are the right line and the curve; and the more comprehensive language of Analysis, in which the quantities considered, whether numerical, concrete, or appertaining to space, are represented by letters of the alphabet. These three forms of language are the basis of classification, and the science of mathematics is divided into three corresponding parts: Arithmetic, Geometry, and Analysis.

The alphabet of the Arithmetical language contains ten characters, called figures, each of which has a name, and when standing by itself indicates as many things as that name denotes. There are but three combinations of these characters—the first is formed by writing them in rows—the second by writing some of them over or under others—and the third, by means of the decimal point. This language, having ten elements and three combinations, is more simple, more minute, and more exact than any other known form of expressing our thoughts. It records all the daily transactions of the world, involving number and quantity. The yearly income—the accumulation of property—the balance sheets of mercantile enterprise are all expressed in numbers, and may be written in figures. These ten little characters are not only the sleepless sentinels of trade and commerce, but they also make known all the practical results of scientific labor.

The language of Geometry is pictorial, and has but two elements, the straight line and curve. The combinations of these simple elements give every form and variety of the geometrical language. Distance, surface, volume and angle, are names denoting portions of space. Under these four names every part of space, in form, extent and dimension, is represented to the mind by means of the straight line and curve. This language is both simple and comprehensive. The shortest distance—the curve of grace and beauty—the smooth surface and the rugged boundary are alike amenable to its laws. It presents to the mind, through the eye, the forms and relative magnitudes of all the heavenly bodies, and, also, of the most minute and delicate objects that are revealed by the microscope. It is the connecting link between theoretical and practical knowledge in the mechanic arts, and the only language in which science speaks to labor. All the works of Architecture, Sculpture and Painting, are but images of the imagination until they assume the geometrical forms.

The language of analysis is more comprehensive than the language of figures or the pictorial language of geometry; indeed, it embraces them both.

Its elements are the leading and final letters of the alphabet, and a few arbitrary signs. The combinations of these elements are few in number and simple in form; and from these humble sources are derived the fruitful language of analytical science. This language is minute, suggestive, certain, general and comprehensive. It will express every property and relation of number—every form which the imagination has given to space—every moment of time which has elapsed since hours began to be numbered—and every motion which has taken place since matter began to move. One or the other of these three forms of mathematical language is in daily use in every part of the world, and especially so in every place where science is employed to guide the hand of labor—to investigate the laws of matter—or to enlarge the boundaries of knowledge.

Of all mysteries, none is greater than the mystery of language. The invisible essence which we call mind, holds no communion with other minds, except through the double system of signs, the language of the eye and the language of the ear. Destroy the power of language, and the lights of knowledge would be extinguished. Man would live only in the present. The past and the future would be equally beyond his reach. Through language we look back over the records of the past, and trace the progress of our race through all its vicissitudes and changes from the very cradle of Creation. The wisdom of philosophy—the power of eloquence—the graces of rhetoric and the inspirations of poetry, thus become the property of every age and the common heritage of mankind. Scientific language reaches even over a wider field. The laws of the material world are the truths which it records, and the thoughts of God, manifested in all the works of the visible creation, are the treasures of its literature.

The first step in mental training is to furnish the mind with clear and distinct ideas, with settled names; each idea and its name being so associated that the one shall always suggest the other. The ideas which make up our knowledge of mathematics fulfill exactly these requirements. They are expressed in a fixed, definite and certain language, which in all its elementary forms may be illustrated by images or pictures, clear and distinct in their outlines, and having names which suggest at once their characteristics and properties.

By means of visible representations of lines, surfaces and volumes, the mind contemplates the abstract, as it were, with a thinking eye. Form, figure, distance, space, and the laws relating to them, are thus rendered familiar through the visibility of pictorial representations. This pictorial language imparts a deep interest, both from its certainty and its influence on the imagination—it attracts and animates the minds of the young, and gradually prepares them for those higher abstractions and mental efforts, of which they are at first incapable.

Most of the errors and conflicts in the Schools of Philosophy have arisen from the double or incomplete sense in which words are employed. The terms are all defined in a common language, but there is no fixed standard beyond the language itself. Each term is viewed from a different stand point, and, like the rainbow painted on the clouds, is different to every spectator, though apparently the same.

Mathematics is free from all such sources of mistake and error. There is no other subject of knowledge in which there is that exact equivalency between the thought and its sign. Number and Space, in all their elementary combinations, may be presented to the mind by pictorial representations. The senses are thus brought to the aid of the conceptive powers, and by means of this double language, the forms, attributes and laws of magnitude, are explained and verified.

The study of mathematics accustoms us to the strict use of this exact and copious language, in which all the terms are exponents of distinct crystallized ideas. Using these terms as instruments of reasoning, we advance with a steady step, secured from the sources or causes of error which are concealed under uncertain or conflicting meanings.

Knowledge is a clear and certain conception of that which is true. Its elements are acquired through the medium of the senses, by observation, experiment and experience; and these three indicate certain relations which the elements bear to each other, and which we express under the general name of law. Law, therefore, is a term of generalization, denoting an order of sequence in phenomena, whether in the material or spiritual, the animate or inanimate world. This order and connection are not obvious to the senses. They are the hidden treasures of knowledge, and are only discovered and brought to light by the highest exercise of the reasoning faculty.

Since the time of Aristotle, the exact law which governs the reasoning faculty has been well known. By careful analysis and a profound generalization, he subjected every principle of deductive reasoning to a single law, expressed by the dictum, and indicated every operation of that law in the syllogism. The system was yet incomplete. The major premise, on which the whole fabric rested, was assumed, not proved. Bacon supplied this deficiency, in showing that all our knowledge rests, ultimately, on the hypothesis of the uniform operation of the laws of nature, and that such uniformity may be inferred by the reasoning faculty, from a collection and comparison of facts, furnished by observation, experiment and experience. This completed the golden circle of logic, and subjected all the laws of nature to the processes of science.

It becomes, therefore, an important inquiry how far the study of mathematics is a means, in the cultivation of the reasoning faculties, through which we derive our scientific knowledge—how far it is a useful gymnastic of the mind—what mental habits it inculcates, and what developments it produces. We have already adverted to its clear, precise, and comprehensive language, and to the elementary ideas, which that language impresses on the mind. Are these ideas isolated—incapable of classification and wanting in the attributes necessary to a logical arrangement?

It is the chief excellence of mathematical science, regarded as a means of mental training, that the definitions and axioms are the prolific sources of every deduction. They are the ultimate premises to which every principle can be referred, and the law of connection which binds together all the truths of this complex system, is the simple law of the syllogism.

Mathematical reasoning, so far as the logic is con cerned, is precisely the same as any other kind of reasoning. It differs from other methods only in the greater preciseness of its language, the nature of the subject and the more obvious relations of the premises to each other, and to the conclusion. It has been urged that these differences are detrimental, rather than useful, in the development of the reasoning faculty—that the exact equivalency between the idea and the language, the fixed and obvious relation of the premises to each other and to the conclusion, leave no scope for originality in the mental processes, and that truth is thus evolved mechanically, rather than intellectually. Another objection has also been found, in the fact that the matter in the mathematical processes is certain, while in all other cases it is contingent—and that to deal with what is certain, in accordance with obvious and fixed laws, disqualifies the mind to deal with what is probable according to laws less obvious and rigorous.

In regard to the second objection, it is quite certain that the degree of probability, in any given case, can only be determined by comparing what is contingent with what is certain—certainty being assumed as the standard—all inferences are relied upon as they approach this standard, and distrusted as they recede from it. Hence, in all systems of intellectual training, having in view the cultivation of the reasoning faculty, the mind should be accustomed to contemplate that which is certain, in order that it may form a true estimate of that which is contingent or probable.

How far the laws which regulate and control the processes of mathematical reasoning are merely mechanical, and how far their study and contemplation confine the mind to a mere routine, is best answered by a careful and searching analysis. The processes begin with obvious and elementary truths, defined by a precise language, and aided, if need be, by pictorial representations. They then advance step by step in a series of regular and dependent gradations, developing the concealed and sublime properties of number and space. These trains of demonstrative reasoning produce the most certain knowledge of which the mind is capable. They establish truth so clearly that none can deny or doubt. The premises are not only certain, but the most obvious truths which can be presented to the mind, and the conclusions result from the most palpable relations of the premises to each other. What discipline can better train the mind to diligence in study—to close and continuous attention—to habits of abstraction—and to a true logical development?

A wide distinction must be made between those processes of mathematics which are merely mechanical and that knowledge of the laws of the science which develops and applies those processes. The calculating machine is a mere instrument, but the discovery and application of the laws of its construction are among the highest efforts of genius. If the machine were dashed to pieces, it could be remodeled, for the law of its construction is known. The conception, therefore, is not mechanical because it is manifested by mechanical agencies. Descartes brought all space within the range and power of analysis, by new methods of representing lines and surfaces. Newton's sublime conception of the law of universal gravitation is developed in the language of Geometry. Does it follow, because the processes of Geometry and the rules for solving equations are reduced to fixed principles and settled methods, that the ubjects to which they may be applied are limited in their nature? or, that the contemplation of these subjects, through this, the only language in which they can be presented to the mind, is likely to give a contracted or one-sided development?

Mathematical Science deals with Number, Space, Time and Motion. Each is a type of the Creator, infinite in itself, and all are under the dominion of universal laws. In the development of these laws, in a language free from obscurity, and in a logic above the influence of passion, sophistry and prejudice, the mind acquires an intensity and ardor which lift it above the strife and petty controversies of earth, into the sphere of the intellectual and absolute. A theorem demonstrated is an indestructible truth; but this is not all, it is connected with antecedent truths of the same kind, and is also a guaranty of our success in new efforts to enlarge the boundaries of knowledge.

In the construction of the mathematical science, we begin with the axiom and proceed from proposition to proposition, under the guidance of a rigorous logic, till we reach the boundaries of that intellectual region which has been already explored. Here we pause, but do not stop; for beyond are hidden truths which excite our innate desire to know, and an ambition and hope of progress. So, when we stretch out the mathematics to explain and embrace the philosophy of the heavens, we proceed from our own planet, in regular gradations, till we reach the remotest orb of our system. Still further on, we enter the region of Arcturus, Orion, the Pleiades and the Milky Way; and, even beyond the smallest star whose light has reached the earth, is unmeasured space, yet perhaps to be surveyed by more perfect instruments, and measured by the known laws of mathematical science.

"There is good room to ask whether the peculiar energy of what might be called the mathematical soul does not carry with it a deep meaning, and declare the truth of man's destination at the first, and of his destiny still to take a place and to act a part in a world of manifested truth and eternal order. Do we venture too far in saying that, when mathematical abstractions of the higher sort take possession of a vigorous reason, there is placed before us a tacit recognition (one among several, all carrying the same meaning,) of the fact that the human mind is so framed as to find its home nowhere but in a sphere within which the absolute and the unchangeable shall stand revealed in the view of the finite intelligence?"*

The term "practical," in its common acceptation, often denotes shorter methods of obtaining results than are indicated by science. It implies a substitution of natural sagacity and "mother wit" for the results of hard study and laborious effort. It implies the use of knowledge before it is acquired—the substitution of the results of mere experiment for the deductions of science, and the placing of empiricism above philosophy. But give to "practical" its true and right signification, and it becomes a word of real

* Isaac Taylor.

import and definite value. In its right sense, it denotes the best means of making the true ideal the actual: that is, of applying the principles of science in all the practical business of life, and of bodying forth in material form the conceptions of taste and genius.

Beyond the obvious application of simple and known principles, the whole problem of the practical lies in the measurement, modification and best uses of the forces of nature. In all the uses and applications of these forces, material substances are employed, and these must be fashioned according to certain forms indicated by scientific formulas. These formulas are constructed from the laws which regulate the cohesion of the particles of the substance employed—the nature of the force to be applied—the amount of that force and the ultimate end to be attained. All these fixed laws of force—all their combinations—and all the forms of the materials employed in using them for practical purposes, can only be reached through the processes and language of mathematics.

Machines and workshops afford marked illustrations of the utility and value of mathematical science, and, in their resolution of difficult practical problems, furnish a striking exhibition of the power of mind over matter. Any one, introduced for the first time to the interior of one of our great factories, would doubtless regard with no small perplexity the equipment and play of so many variously directed instruments of motion—the great size and extent of the whole structure—the jar which startles at first, but by the steadiness of its pulsations soon persuades you to take the cadence and measure of the great machine, and to appropriate, as it were, a share of the producing power—and it would be strange if you were not also persuaded that all this bewildering procession of complex returning movements must be under the guidance of some great scientific law.

All the parts of that complicated machinery are adjusted to each other, and were indeed so arranged, according to a given plan, before a single wheel was formed by the hand of the forger. The power necessary to do the entire work was first carefully calculated, and then distributed throughout the ramifications of the machinery. Each part was so arranged as to fulfill its office. Every circumference and band and cog, has its specific duty assigned to it. They are connecting parts of an entire practical scientific system, over which one of the parts, fitly called the governor, is most ingeniously appointed to preside. It is the function of this apt and beautiful contrivance to regulate the force which shall drive the whole, according to a uniform speed; and it performs the office with such sensibility and seeming intelligence, that, on the slightest increase of velocity, it commences and executes, with easy gradations, a diminution of the moving force of the machine, and as instinctively calls up additional power the moment that the speed slackens. All this is the result of calculation. When the curious shall visit these exhibitions of ingenuity and skill, let them not suppose that they are the offspring of chance and experiment. They are the embodiment, by intelligent labor, of the results of the most difficult investigations of science.

The Steamship affords another impressive illustration of theoretical and practical science. Observe her form—how perfect in all its parts—how beautiful in outline—how exact in proportion. See how gracefully she rests upon the water, which she scarcely seems to touch. On the upper deck, the masts and ropes, the yards, the spars, the booms and sails, are all adjusted to the proper angle and are the instruments by which the power of the wind is pressed into the service of commerce. But this is not the power on which she relies. The great mechanical contrivance, to which I have alluded, which just now shook the earth with its jar, is to be readjusted and folded within a structure having its own peculiar form and limits, designed for special functions and moving on a new element. The source of power is a simple change in the form of a fluid.

The massive cylinders, the huge levers, the lifting and closing valves are contrivances to convey this power to the water wheels, where the resistance of the water, according to known laws, transfers it to the ship itself.

Over all this complication of machinery—over all this variety of principle and workmanship, science has waved her magic wand. There is not a cylinder whose dimensions were not measured—not a lever whose power was not calculated, nor a valve which does not open and shut at the appointed moment. There is not, in all this structure, a bolt, a screw, or rod which was not provided for before the great shaft was forged, and which does not bear to that shaft a proper proportion.

The language of Geometry and Number furnished the architect with all the signs and instruments of thought necessary to a perfect ideal of his work, before he took the first step in its execution. It also enabled him, by drawings and figures, so to direct the hand of labor as to form the actual after its pattern—the ideal. The various parts may be constructed by different mechanics, at different places, but the law of science is so certain that every part will have its right dimensions, and when all are put together they form a perfect whole.

When the work is done and the ship takes her

departure for another continent, a small piece of iron, a few inches in length, poised on its centre, under the influence of a known force, is the little pilot which guides her over trackless waters. Science has also provided, for daily use, maps and charts of the port which she leaves, of the ocean to be traversed and of the coasts and harbors which are to be visited. On these are marked the results of much careful labor. The shoals, the channels, the points of danger and the places of security, are all indicated. Near by hangs the Barometer, constructed from mathematical formulas, to indicate changes in the weight of the atmosphere and give warning of the approaching tempest. In close proximity are the Sextant and the Tables of Bowditch. These are the simple contrivances which science has furnished to correct the errors of the needle, by observations on the heavenly bodies, and to determine the exact position of the vessel at any moment of the voyage. Thus, practical science, which determined the form of the vessel best adapted to a given velocity, which measured and distributed the propelling force and which guided the hand of the mechanic in every workshop, is, under Providence, the means of conducting her in safety over the ocean. It is, indeed, the cloud by day and the pillar of fire by night.

The construction of railways is a recent and most important application of science. The mechanic arts, commerce and civilization have all received an impulse in this new development of power. The chariots of commerce, which rush with such dizzying velocity over the iron bands which now nearly encircle the globe, are all guided by immutable laws that have been carefully developed by the aid of diagrams and equations. When you see the long train, with its locomotive, ascending the mountain, fear not, for science traced the curve and balanced the forces. When the mountain is to be pierced instead of being scaled, a few lines drawn on paper indicate the precise points, at the opposite extremities, where the work is to be begun; and after years of labor the two working parties meet near the centre, and in the exact line established before the ground was broken.

In every case where power is employed, either to produce motion or to maintain a state of rest, the mechanical principle of force and resistance must be considered and discussed. Mathematics is the only form of language which connects science with all the mechanic arts and guides the hand of labor as it bodies forth the conceptions of the mind. It is, therefore, the only true basis of the practical; and perhaps it is not too much to add, that whatever is true and just in the practical is the actual of an antecedent ideal.

Material objects are the first things which attract our notice. We behold the earth filled with products and teeming with life. We note the return of day and night at regular intervals—the coming of summer and winter, and the succession of heat and cold. We see the sun in the firmament—we turn our eyes to the starry heavens and behold the sentinels of night as they look down upon us. Facts, often observed, suggest the idea of causes—and, when science scatters her light over the pathway of the past and the future, we learn the existence of general laws imparted by the fiat of Him who created all things—and come to understand that mind in all its attributes, and matter in all its forms, are subject to those laws—and that their study is the noblest employment of our intellectual nature.

To the uneducated man, all the world is a mystery. He does not see how so great a uniformity can exist with the infinite variety which pervades every department of nature, animate and inanimate. In the animal kingdom no two of a species are exactly alike; and yet the general resemblance and conformity are so close that the Naturalist, from the examination of a single bone, finds no difficulty in determining the species, size and structure of the animal. So, also, in the vegetable and mineral kingdoms, where all the structures of growth and formation, though infinitely varied, are yet conformable to like general laws.

The wonderful mechanism displayed in the structure of animals was but imperfectly understood, until analyzed and illustrated by the principles of science. Then, a general law, applicable to every case involving power and motion, was found to pervade the whole. Every bone is proved to be of that length and diameter best adapted to its use—every muscle is inserted at the right point, and works about the right centre—the feathers of every bird are shaped in the best form, and the curves in which they cleave the air are the best adapted to velocity. It is demonstrated, that in every case, and in all the varieties of form, in which forces are applied, either to increase power or gain velocity, general laws have been established to produce the desired results. Thus science makes known to us the foreknowledge and wisdom of the Creator.

But inanimate nature also speaks to us in the language of general laws, and it is in the investigation and interpretation of these laws that mathematical science finds its widest range and its most striking applications. Experience, aided by observation and enlightened by experiment, is the recognized fountain of all knowledge of nature. On this foundation Bacon rested his philosophy. He saw that the deductive process of Aristotle, in which the conclusion does not reach beyond the premises, was not progressive. It might, indeed, improve the reasoning process, cultivate habits of nice discrimination and give great proficiency in verbal dialectics; but the basis was too narrow for that expansive philosophy which was to unfold and harmonize all the laws of nature. Hence, he suggested a careful examination of nature in every department, and thus laid the foundations of a new philosophy. Nature was to be interrogated by experiment; observation was to note the results and gather the facts into the store-house of knowledge. Facts, so obtained, were subjected to analysis and collation, and from such classification general laws were inferred, by a reasoning process called Induction.

This new philosophy gave a startling impulse to the mind, and to knowledge. Its subject was nature —material and immaterial; its object, the discovery and analysis of those general laws which pervade, regulate and impart uniformity to all things; its processes, experience, experiment and observation for the ascertainment of facts, analysis and comparison for their classification, and the reasoning process for the establishment of general laws. But the work would have been incomplete without the aid of deductive Science. General laws, deduced from many separate cases, by induction, needed additional proof; for they might have been inferred from resemblances too slight, or from coincidences too few. Mathematics affords such proofs.

Every branch of natural philosophy was originally experimental; each generalization rested on a special induction, and was derived from its own distinct set of observations and experiments. From being sciences of pure experiment, or sciences in which the reasonings consist of no more than one step, and that step an induction, all these sciences have become, to some extent, and some of them in nearly their whole extent, sciences of pure reasoning: thus, multitudes of truths, already known by induction, from as many different sets of experiments, have come to be exhibited as deductions, or corollaries from inductive propositions of a simple and more universal character. Thus, Mechanics, Acoustics, Optics and Chemistry, have successively been rendered mathematical: and Astronomy was brought by Newton within the laws of general mechanics.

The substitution of this circuitous mode of proceeding, for a process apparently much easier and more natural, is held, and justly too, to be the greatest triumph in the investigation of nature. But it is necessary to remark that although, by this progressive transformation, all sciences tend to become more and more deductive, they are not, therefore, the less inductive: for every step in the deduction rests on antecedent induction.*

We can now, therefore, perceive what is the generic distinction between sciences which can be made deductive, and those which must, as yet, remain experimental. The difference consists in our having been able, in the first case, and not in the second, to establish a set of first inductions, from which, as from a general law, we are able to draw a series of connected and dependent truths. For example, when Newton, by observing and comparing the motions of several of the bodies of the solar system, discovered that each, whether its motions were regular or apparently anomalous, conformed to the law of moving around a common centre, urged by a centripetal force, varying directly as the mass and inversely as the square of the distance, he inferred the existence of the law for all bodies; and then demonstrated, by the aid of mathematics, that no other law could produce such motions. This is the most striking example which has yet occurred of the transformation, at a single stroke, of a science, which was in

* Mill's Logic.

a great degree experimental, into one purely deductive.

It is in the great problem of the solar system that mathematical science displays its omnipotent power. The sun himself, manifesting his inexpressible glory by the floods of golden light which he scatters through the immensity of space, is yet subjected to the analytical formula, and must confess to it, from his more than imperial throne—his exact dimensions—his weight and balancing power, and his relative importance when compared with the smallest mote which his own light has revealed. It is thus that the intellectual power, aided and stimulated by the processes of mathematical science, has been able to trace backwards, to the earliest past, all the motions of the heavenly bodies and to bring the remotest future of the planetary system within the range of its computations. It is thus that man, inhabiting one of the smallest planets of the system, computes the celestial cycles and determines all the laws of the movement of the celestial machinery.

He has done even more than this. Those vagrant bodies of the heavens which occasionally visit our system, and which seem to have escaped from their own spheres and to wander heedlessly through space, are yet subjected to the power of analysis. A few observations, made by the practical astronomer, afford the necessary elements for computing the forms of their orbits and their periodic times; and in distant years, at the indicated moment, the comet again blazes in the sky. In short, before this august power all nature yields up the mystery of her laws. If, then, we would enter her spacious temple, and seek after the knowledge which is there, let us not forget the Aladdin's lamp of mathematical science, which, being properly touched, will disclose more treasures than have ever been described in Eastern fable.

The place which mathematics should occupy in a system of collegiate instruction is an inquiry of the gravest import, and necessarily involves the question, What should be the nature of the system itself?

It was stated, in the opening address, on the highest authority, "that the end of a liberal education is the general and harmonious evolution of all the faculties and capacities of the mind in their relative subordination." It is not the base, nor the massive shaft, nor the beautiful forms of the capital, which fill the mind as we gaze on the Corinthian column; but it is their unity and the general effect of their combination. It is the whole mind, in all its intellectual and emotional faculties, to which the experienced educator addresses himself.

So far as our knowledge extends, we have found

in that mysterious essence, the mind, a faculty adapted to the apprehension of every law, and an emotion corresponding to the contemplation of every object. May not the reverse of this proposition be true? May it not be, that for every faculty of the mind, whether intellectual or emotional, there exists, somewhere, a proper object of contemplation? and that the perfection of our knowledge and being will be attained when all such objects are found? It is in accordance with this law that different studies cultivate different powers of the mind, and that it requires the study of many subjects to give a general and harmonious evolution of all its faculties. Mathematics does not equally cultivate every faculty—it is the massive trunk and outward form, but language, literature, and moral culture, are the sap which ascends within, and which is necessary to give beauty to the foliage and health and harmony to the whole development. All the colors of the rainbow, which are painted on the clouds, are necessary to the perfect light of day—so every light of knowledge is required in the perfect illumination of the mind.

It is the special function of mathematical studies, to cultivate the faculty of abstraction and the habit of intense and continued attention—to establish in the mind a self-centering power that shall subordinate all the intellectual faculties to the control of the will

—to create, as it were, a governor of the intellectual machinery, that will give harmony and uniformity to all its motions. As an elementary formula of logic, it is the most simple and perfect. As a drill, in the structure and use of language, in its primary forms, no exercise insures greater precision in the use of words, or imparts to the mind as certain relations between the signs and the things signified. In its higher branches, it is even an aid in the study of theology; for it constantly raises the mind to the contemplation of the Unchangeable and the Infinite. Mathematics, therefore, is an aid and auxiliary in every other branch of study. It may be pursued too exclusively—the mind may become too much absorbed by its machinery and formulas; but this danger is common to the study of every other subject. A life spent exclusively on the Greek Grammar would not make a Greek scholar; nor can the wide field of deductive reasoning be explored by repeating the formulas of the dictum and syllogism.

Concurring fully in what was said, in the opening address, concerning the great value of the study of the Greek and Latin languages, and also in the merited eulogium of the manner in which these languages are taught in this institution, I may yet be permitted to say, that there is another language far more comprehensive than either or both of them:

the language of mathematics, which embraces within its ample folds all the laws of the material universe. This language takes us back to the birth of matter, and measures and records every step which each planet has taken since it began to move. Yea, more: it is prophetic—it reveals all future motions, and indicates the precise places which all matter must occupy, at any given instant of future time.

This is the language in which the practical astronomer studies the heavens. It is the telegraphic wire which has enabled him to communicate with every planet of our system—to measure its diameter, its specific gravity, the dimensions of its orbit, its times of revolution and its balancing power in the system of the universe. It is this language which has enabled him to bring the ring of Saturn into his own study, where he sees it face to face, and, as it were, touches the very particles of matter of which it is composed.

This language has enabled the naturalist to trace the dominion of law over all matter endowed with life. The contemplation of the minute objects of creation may appear, at first sight, unworthy the labors of the highest genius—but it is quite otherwise. The turtle's egg, the little gnat whose tiny wings vibrate five hundred times in a second, and the entire solar system, are each an embodiment of a thought of God. Whether we look through the microscope or the telescope, we are equally instructed in the wonders of creative power and universal law.

But science is not all in all. It does not compass the final aim and ultimate end of our being. Though it reaches back to the time when God said "Let there be light and there was light," and forward to the time when "there shall be a new heaven and a new earth"—though it measures all space—though it explains all laws relating to matter and motion—though it transports us to the central point of the physical universe, whence we behold the heavenly hosts moving in celestial harmony: yet, when we approach that mysterious line where the finite terminates and the infinite begins, new visions open to the mind—all science and human knowledge fade away like castellated clouds made brilliant by the setting sun—Faith then arises in supernal beauty, and, with veiled eyes and trembling voice, we confess, "In the beginning was the Word, and the Word was with God, and the Word was God."

Made in the USA
Middletown, DE
29 November 2022

16456953R00022